SPACE WONDERS

DWARF PLANETS

BLACK RABBIT BOOKS

TABLE OF CONTENTS

1. Dwarf Planets 4
2. Pluto 6
3. Eris 10
4. Makemake 13
5. Ceres 16
6. Haumea 19
More to Explore 22

1

Dwarf Planets

Not all planets are equal. Some are called dwarf planets. Like other planets, they are round. They **orbit** a star. But they are too small to be a planet. They often have comets and other planets nearby. Big planets do not.

NASA has named five dwarf planets. They are all in our **solar system**. There may be more.

Did You Know?
All dwarf planets are smaller than Earth's Moon.

2

Pluto

Pluto (PLOO-toh) was found in 1930. It used to be our solar system's ninth planet. But it was too small. It could not push other space objects away. In 2006, NASA named Pluto a dwarf planet. It has five moons. The largest is named Charon.

Pluto is about 3.7 billion miles (6 billion kilometers) from Earth. Its **atmosphere** is made of deadly gases. The temperature is hundreds of degrees below zero. Only one spacecraft has studied Pluto. NASA's *New Horizons* flew past it in 2015.

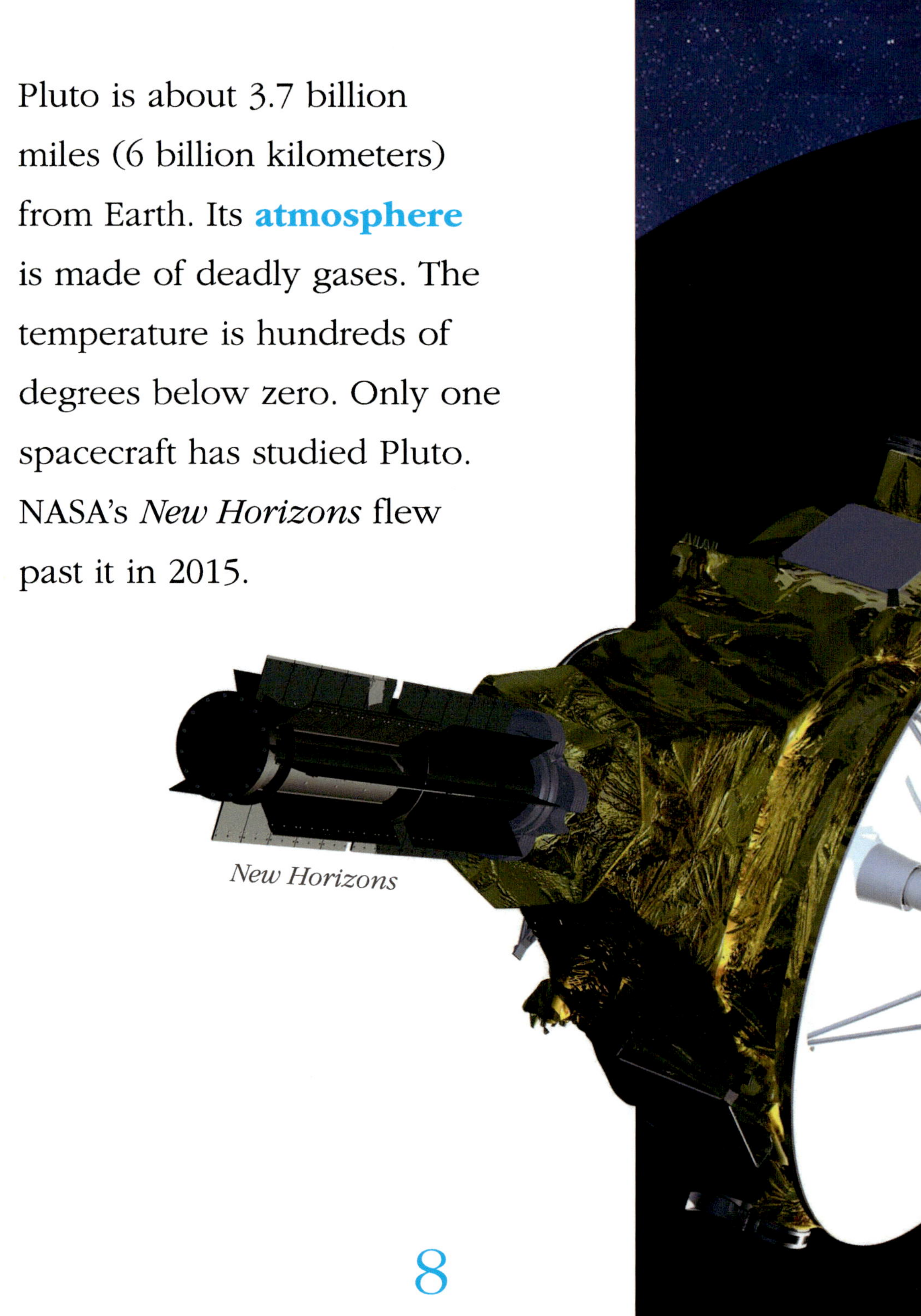

New Horizons

Think About It

Many people were upset when Pluto was named a dwarf planet. Why do you think that is?

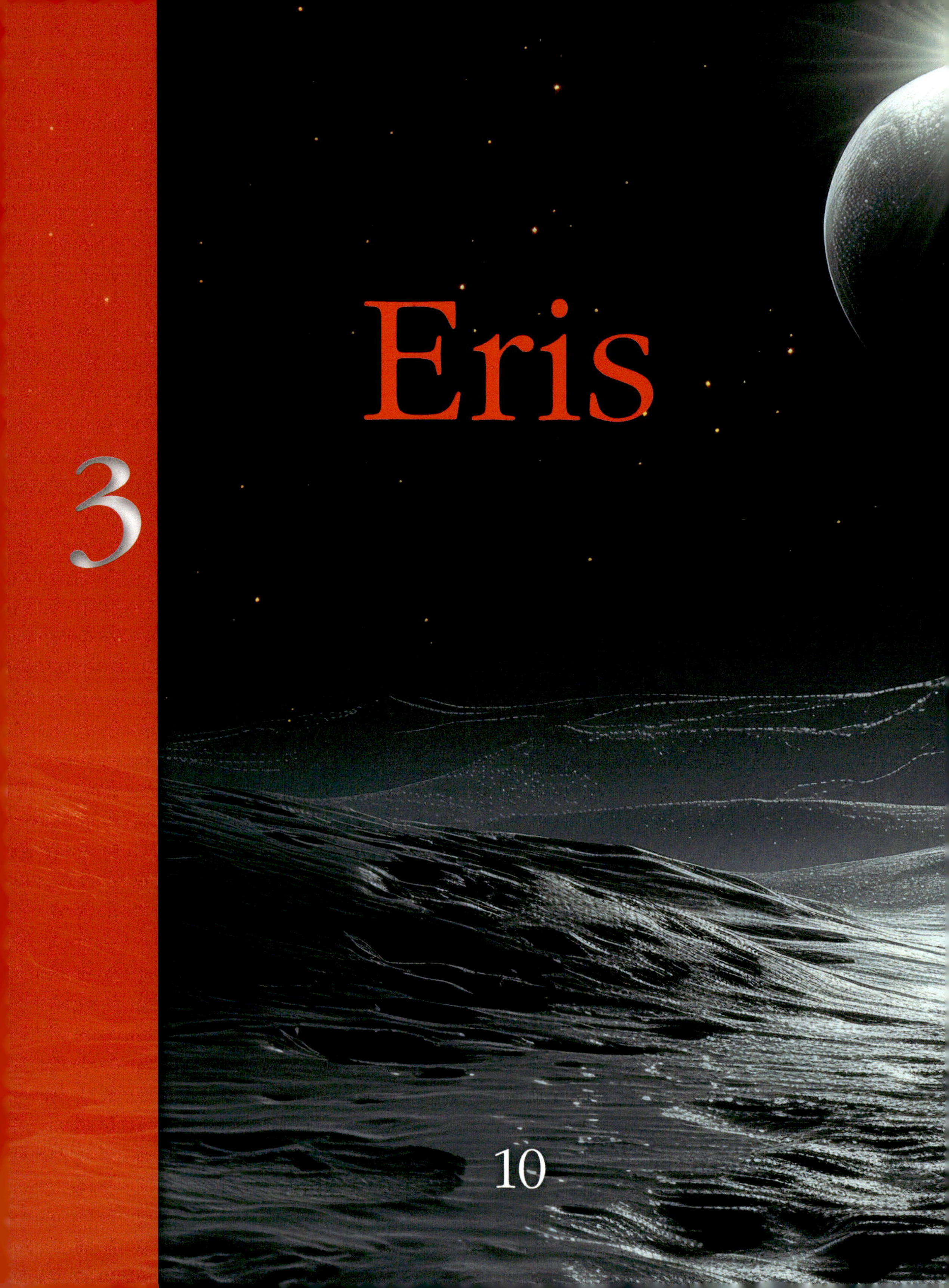

3

Eris

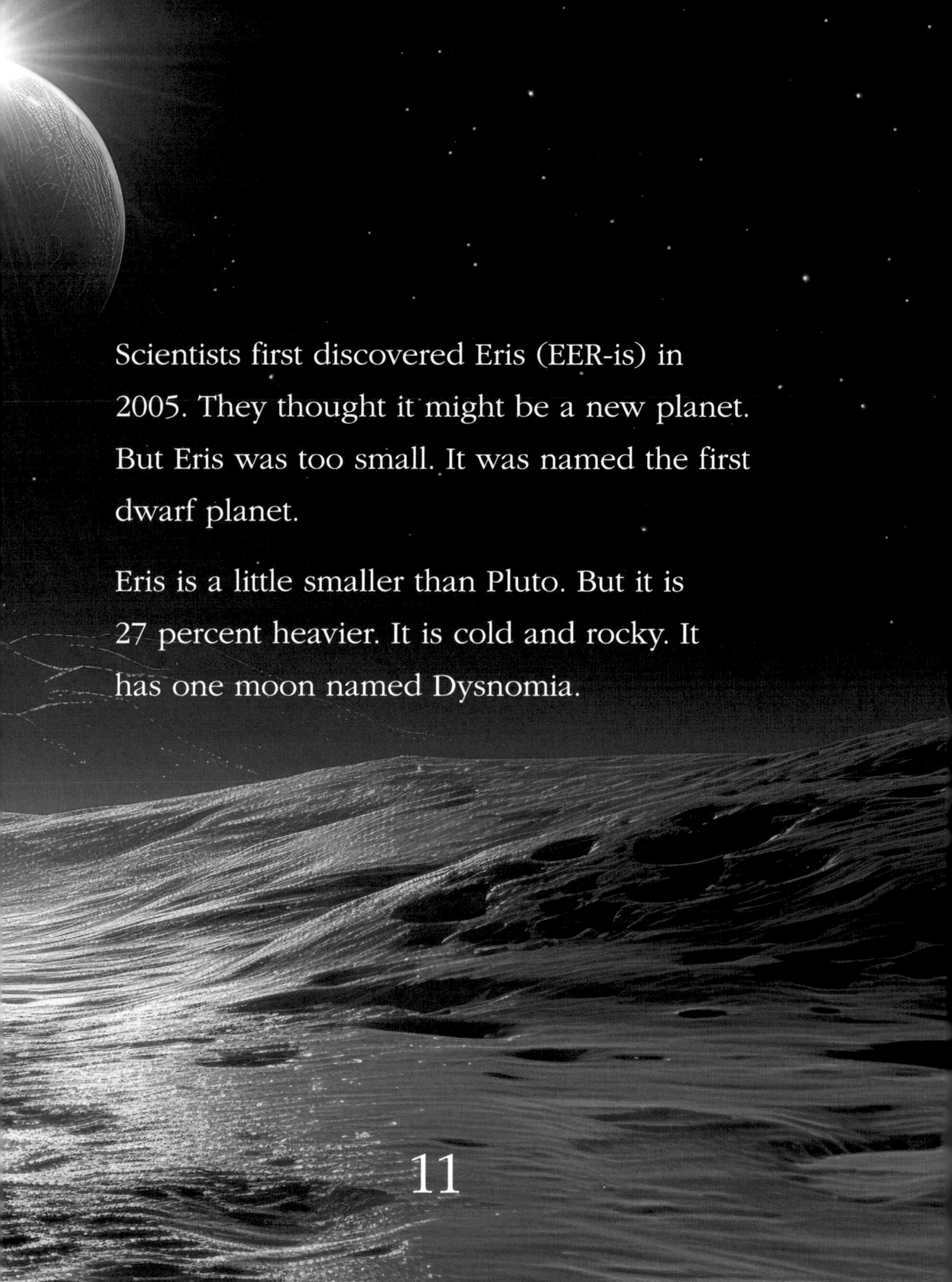

Scientists first discovered Eris (EER-is) in 2005. They thought it might be a new planet. But Eris was too small. It was named the first dwarf planet.

Eris is a little smaller than Pluto. But it is 27 percent heavier. It is cold and rocky. It has one moon named Dysnomia.

Scientists do not know much about Eris. It is far from Earth. It is in the Kuiper Belt. This is past Neptune. Many dwarf planets and icy objects are here.

Did You Know?
Eris is named after a Greek goddess.

4

Makemake

The year 2005 was a big one for dwarf planets. Scientists found Eris that year. They also found Makemake (MAH-kee-mah-kee). Like Eris and Pluto, this dwarf is far from Earth. It is also in the Kuiper Belt.

Makemake is a cold world. No life could ever survive there. Its surface is hard to see. But scientists know it is red-brown in color.

Did You Know?
Makemake was first called Easterbunny.

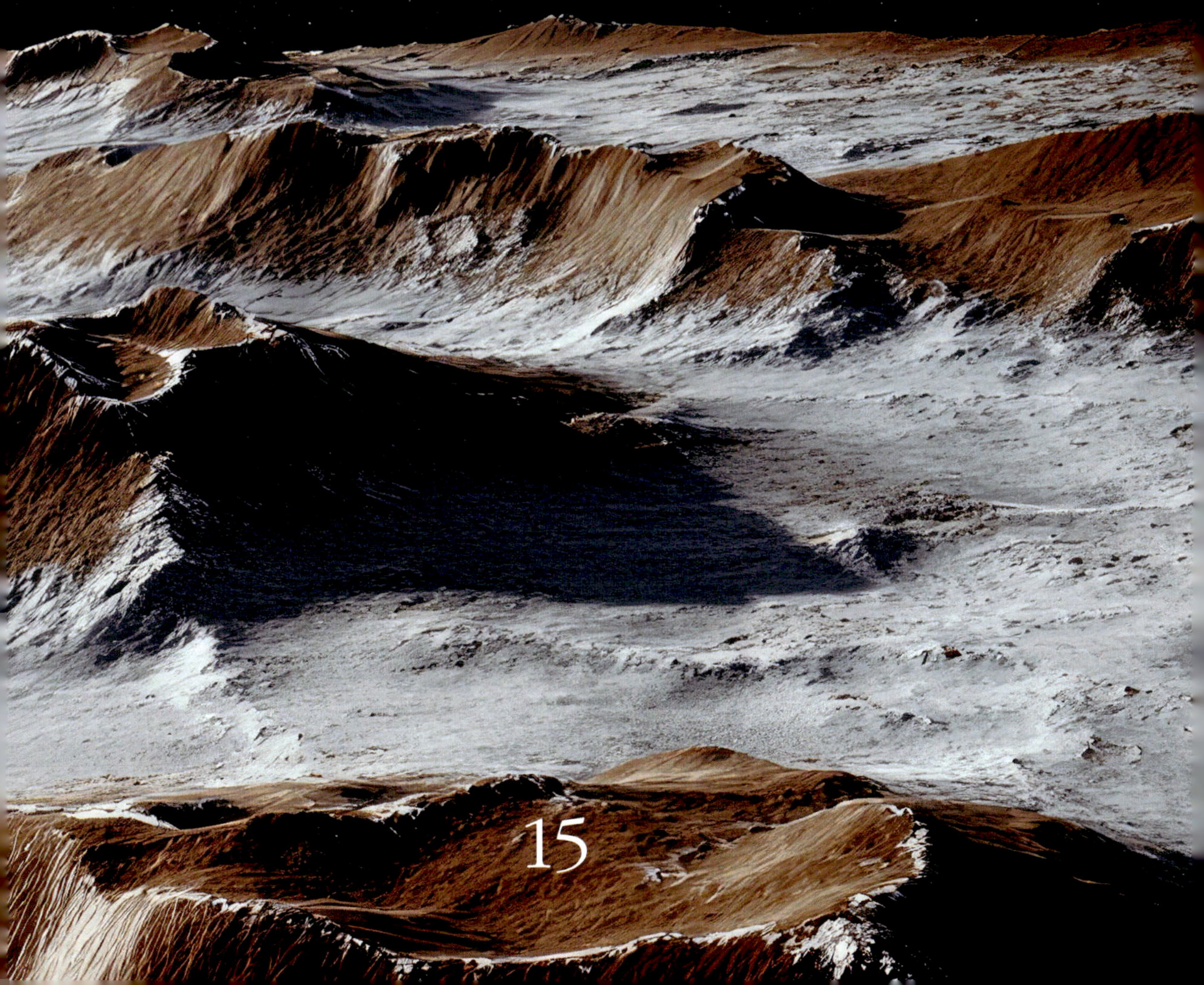

5

Ceres

Most dwarf planets are far away from Earth. Not Ceres (SEER-eez)! It is in the **asteroid** belt. This is between Mars and Jupiter.

Scientists thought Ceres was an asteroid. Then they studied it more closely. In 2006, they named it a dwarf planet.

The NASA spacecraft *Dawn* explored Ceres. It found water **vapor** and salt. Could Ceres support life? Scientists are studying the dwarf to find out.

Think About It

What do planets need to support life?

Haumea

Most planets are round. But Haumea (HAH-may-uh) is not. This odd dwarf spins quickly. Its days are four hours long! It cannot keep a round shape. Instead, it looks like an egg.

Scientists found Haumea in 2003. The dwarf planet has two moons. It also has at least one ring. Haumea is 4.7 billion miles (7.6 billion km) from Earth. Scientists hope to learn more about it in the future.

Think About It

How do scientists study objects that are far away from Earth?

MORE TO EXPLORE

FANTASTIC FACTS

A day on Makemake is about as long as a day on Earth or Mars.

Ceres does not tilt as it spins. This means the dwarf planet has no seasons.

Sometimes Eris's atmosphere collapses and freezes. When that happens, it snows!

Pluto's tilted orbit crosses Neptune's path. Sometimes, it is closer to the sun than our eighth planet.

Haumea's ring is made of tiny pieces of ice.

Ceres might have a hidden ocean under its layer of ice.

MORE TO EXPLORE

COOL COMPARISONS

How big are the dwarf planets?
If Earth was the size of a nickel, then:

Pluto and **Eris** would each be the size of a popcorn kernel.

Ceres would be the size of a poppy seed.

Makemake would be the size of a mustard seed.

Haumea would be the size of a sesame seed.

MORE TO EXPLORE

RESOURCES

Glossary

asteroid (AS-tuh-roid) A small rocky body that circles the sun.

atmosphere (AT-muhs-feer) The gases that surround a planet.

solar system (SOH-luhr SYS-tum) The planets and space objects that circle the sun.

orbit (AWR-bit) The path taken by one body circling around another body.

vapor (VEY-per) A substance in the form of a gas.

Read More

Betts, Bruce. *Dwarf Planets: Small Round Worlds.* Minneapolis: Lerner Publications, 2025.

Rathburn, Betsy. *The Dwarf Planets.* Minneapolis: Bellwether Media, 2023.

Index

asteroid belts, 17
atmosphere, 8, 17, 22
dwarf planet rules, 5, 7, 11
Kuiper Belt, 12, 14
moons, 7, 11, 20
NASA, 5, 7, 8, 17
rings, 20, 22
shape, 5, 20
spacecraft, 8, 17

TOP RANK is published by Black Rabbit Books, P.O. Box 227, Mankato, MN, 56002. •

Top Rank is an imprint of Black Rabbit Books • Series designe by Danny Nanos • Book designed by Jason Knudson • Photographs © Adobe Stock/BreizhAtao, 19, Nonema, 10–11, 18, Pannin, 13, 14–15, 16–17 SN, 20–21; Freepik/anastasil7, 23, Kavalenkava, 23, macrovector, 23, rawpixel.com, cover, 4–5, sonulkaster, 23, valadzionak_volha, 23, wirestock, 2–3 7; NASA/Johns Hopkins University Applied Physics Laboratory/Southwest Research Institute, 7, 8–9; NASA/JPL/USGS, cover, 5; NASA/JPL-Caltech, 12 NASA/JPL-Caltech/UCLA/MPS/DLR/IDA, cover, 5; Shutterstock/Dotted Yeti, 6, Sudakarn Vivatvanichkul, cover, 2, 4–5, 3dector, cover, 5; Wikimedia Commons Altes Museum, 12, ESO/L. Calçada and Nick Risinger (skysurvey.org), cover, 4, NASA/Apollo, 17 crew, 23, NASA/Johns Hopkins University Applied Physic Laboratory/Southwest Research Institute, cover, 4 • Printed in the United States of America

Library of Congress Cataloging-in-Publication Data: Names: Mattern, Joanne, 1963- author. | Title: Dwarf planets / Joanne Mattern. | Description: Mankat MN: Black Rabbit Books, [2026] | Series: Space wonders | Includes bibliographical references and index. | Audience: Ages 8–11 | Audience: Grades 2–3 Identifiers: LCCN 2025011964 (print) | LCCN 2025011965 (ebook) | ISBN 9781645825081 (library binding) | ISBN 9781645825265 (paperback) | ISB 9781645825449 (ebook) | Subjects: LCSH: Dwarf planets—Juvenile literature. | Classification: LCC QB698 .M38 2026 (print) | LCC QB698 (ebook) | DD 523.44—dc23/eng/20250324 | LC record available at https://lccn.loc.gov/2025011964 | LC ebook record available at https://lccn.loc.gov/2025011965